"Extincton is Greed Driven", is an update of "The World According to Vern Book 4".

I0481889

Contents

We are nearing the end of the year 2017 and we are no closer to resolving Nuclear Threats by North Korea!

The Road of life is full of Ruts! Take the Buck and the Doe as a start. The Buck snorts as he trots behind the Doe following her Scent. His hormones are running wild. Finally, he catches up to her, nuzzles her and she stands still as he mounts her and after a brief but traumatic entry in her womb he dismounts exhausted and stumbling away he go's in pursuit of another doe. She is left to carry the gift of his sperm!

Blue Sky

My neighbor Ted Fry recently posted sunrise pictures! The sky appears beautiful. See the cover of this book!

"Blue Skies and Red Sunsets

- The Electromagnetic and Visible Spectra
- Visible Light and the Eye's Response
- Light Absorption, Reflection, and Transmission
- Color Addition
- Color Subtraction
- Blue Skies and Red Sunsets
  The sun emits light waves with a range of frequencies. Some of these frequencies fall within the visible light spectrum and thus are detectable by the human eye. Since sunlight consists of light with the range of visible light frequencies, it appears white. This white light is incident towards Earth and illuminates both our outdoor world and the atmosphere that surrounds our planet.

As discussed earlier in Lesson 2, the interaction of visible light with matter will often result in the absorption of specific frequencies of light. The frequencies of visible light that are not absorbed are either transmitted (by transparent materials) or reflected (by opaque materials). As we sight at various objects in our surroundings, the color that we perceive is dependent upon the color(s) of light that are reflected or transmitted by those objects to our eyes. So, if we consider a green leaf on a tree, the atoms of the chlorophyll molecules in the leaf are absorbing most of the frequencies of visible light (except for green) and reflecting the green light to our eyes. The leaf thus appears green. And as we view the black asphalt street, the atoms of the asphalt are absorbing all the frequencies of visible light and no light is reflected to our eyes. The asphalt street thus appears black (the absence of color).

In this manner, the interaction of sunlight with matter contributes to the color appearance of our surrounding world. In this part of Lesson 2, we will focus on the interaction of sunlight with atmospheric particles to produce blue skies and red sunsets. We will attempt to answer these two questions:

. Why are the skies blue?
. Why are the sunsets red?

Visible Light Spectrum

R O   B I **V**

long wavelength   short wavelength
low frequency   high frequency

Least readily scattered   Most readily scattered
part of spectrum.   part of spectrum.

Why are the skies blue?

The interaction of sunlight with matter can result in one of three wave behaviors: absorption, transmission, and reflection. The atmosphere is a gaseous sea that contains a variety of types of particles; the two most common types of matter present in the atmosphere are gaseous nitrogen and oxygen. These particles are most effective in scattering the higher frequency and shorter wavelength portions of the visible light spectrum. This scattering process involves the absorption of a light wave by an atom followed by reemission of a light wave in a variety of directions. The amount of multidirectional scattering that occurs is dependent upon the frequency of the light. (In fact, it varies according to f^4.) Atmospheric nitrogen and oxygen scatter violet light most easily, followed by blue light, green light, etc. So as white light (ROYGBIV) from the sun passes through our atmosphere, the high frequencies (BIV) become scattered by atmospheric particles while the lower frequencies (ROY) are most likely to pass through the atmosphere without a significant alteration in their direction. This scattering of the higher frequencies of light illuminates the skies with light on the BIV end of the visible spectrum. Compared to blue light, violet light is most easily scattered by atmospheric particles. However, our eyes are more sensitive to light with blue frequencies. Thus, we view the skies as being blue in color.

Why are sunsets red?

Meanwhile, the light that is not scattered is able to pass through our atmosphere and reach our eyes in a rather non-interrupted path. The lower frequencies of sunlight (ROY) tend to reach our eyes as we sight directly at the sun during midday. While sunlight consists of the entire range of frequencies of visible light, not all frequencies are equally intense. In fact, sunlight tends to be most rich with yellow light frequencies. For these reasons, the sun appears yellow during midday due to the direct passage of dominant amounts of yellow frequencies through our atmosphere and to our eyes.

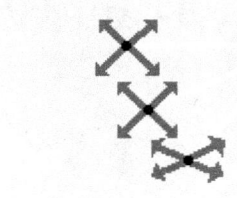

The yellow appearance of the noon-day sun is due to the scattering of the higher frequencies of sunlight.

The appearance of the sun changes with the time of day. While it may be yellow during midday, it is often found to gradually turn color as it approaches sunset. This can be explained by light scattering. As the sun approaches the horizon line, sunlight must traverse a greater distance through our atmosphere; this is demonstrated in the diagram below.

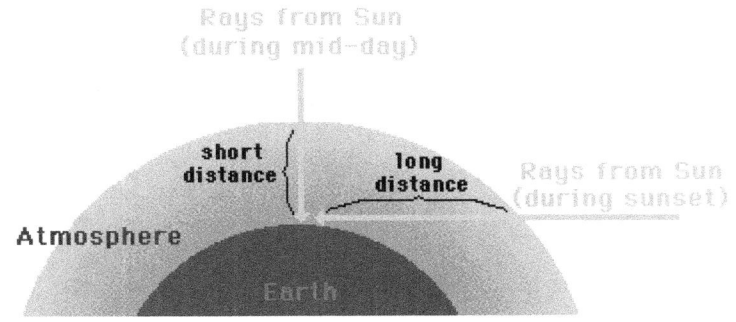

As the path that sunlight takes through our atmosphere increases in length, ROYGBIV encounters more and more atmospheric particles. This results in the scattering of greater and greater amounts of yellow light. During sunset hours, the light passing through our atmosphere to our eyes tends to be most concentrated with red and orange frequencies of light. For this reason, the sunsets have a reddish-orange hue. The effect of a red sunset becomes more pronounced if the atmosphere contains more and more particles. The presence of sulfur aerosols (emitted as an industrial pollutant and by volcanic activity) in our atmosphere contributes to some magnificent sunsets (and some very serious environmental problems).

The Wonders of Physics

Photograph of Maui sunset by Becky Henderson"

(Source: The Physics Classroom Light Waves and Color - Lesson 2 - Color and Vision)

Are we so easily duped? The beautiful colors can be caused by pollutants.

HOWEVER, it appears that Volcanic emissions worldwide are likely responsible for red skies in November 2017.

New activity/unrest was reported for 2 volcanoes between November 22 and 28, 2017. During the same period, ongoing activity was reported for 12 volcanoes.

New activity/unrest: Agung, Bali (Indonesia) | Great Sitkin, Andreanof Islands (USA).

Ongoing activity: Aira, Kyushu (Japan) | Aoba, Vanuatu | Cleveland, Chuginadak Island (USA) | Copahue, Central Chile-Argentina border | Dukono, Halmahera (Indonesia) | Ebeko, Paramushir Island (Russia) | Kilauea, Hawaiian Islands (USA) | Popocatepetl, Mexico | Reventador, Ecuador | Sabancaya, Peru | Sheveluch, Central Kamchatka (Russia) | Sinabung, Indonesia.

President Trump has removed emission standards for factories, more fake news! The only fake is a President with No Clothes!

God gave us Science as A Gift! Yes, Pseudo Scientists Abuse This Gift. Paid Scientists for the Tobacco Industry claimed for years that Cigarettes did not cause cancer! Only recently has the Tobacco industry admitted that Cigarettes Cause Cancer! How long are we going to allow a President with no clothes rule our country?

United or Divided

President Trump expounded during the presidential campaign and numerous times since that he wants a united USA and world governments to work together! He also claims that he was against a second invasion of Iraq. Trump used his alleged opposition to the invasion as a bludgeon in the Republican primary, tearing down Jeb Bush and positioning himself as a foreign policy thinker.

However, there is no evidence that he voiced opposition to the war!

Validated statements by Trump:

"I'm no warmonger. But the fact is, if we decide a strike against Iraq is necessary, it is madness not to carry the mission to its conclusion. When we don't we have the worst of all worlds: Iraq remains a threat, and now has more incentive than ever to attack us. (Trump'2000 book, The America We Deserve)

"Look at the war in Iraq and the mess that we're in. I would never have handled it that way. Does anybody really believe that Iraq is going to be a wonderful democracy where people are going to run down to the voting box and gently put in their ballot and the winner is happily going to step up to lead the country?

C'mon. Two minutes after we leave, there's going to be a revolution, and the meanest, toughest, smartest, most vicious guy will take over. And he'll have weapons of mass destruction, which Sadam didn't have."
(Esquire,2004)

"How do they get out? You know how they get out? They get out. That's how they get out. Declare victory and leave. Because I'll tell you, this country is just going to get further bogged down They're in a civil war over there, Wolf. There's nothing that we're going to be able to do with a civil war. They are in a major civil war." (CNN, 2007)

Trump phrases from 2017 UN Speech:

"In foreign affairs, we are renewing this founding principle of sovereignty. Our government's first duty is to its people, to our citizens, to serve their needs, to ensure their safety, to preserve their rights and to defend their values. "Trump said. "I will always put America first, just like you, as the leaders of your countries will always and should always put your countries first."

Trump also said that sovereignty can be a "call for action," noting that" all people deserve a government that cares for their safety, their interests and their well-being."

In vowing that "we will stop radical Islamic terrorism, "Trump dimmed hope that he had begun to understand the damaging impact the controversial phrase can have on relations with Muslims in the US and abroad.

End of 2017

Following successful visits in Saudi Arabia on May 20,2017; Israel, Japan, South Korea, Japan,

China, Pope Francs in Italy, Great Britain and ending with the G-7 nations in Sicily May 27.

Trump returns to the US to send out a False Anti-Muslim Tweet! President Trump retweeted a video that purported to show a "Muslim migrant" beating up "a Dutch Boy on crutches." But, according to the Netherlands Embassy in the United States, the attacker wasn't an immigrant. He was born and raised in the Netherlands.

The embassy chastised the U.S. president for spreading false information. "Facts do matter," it said in a statement on Twitter hours after the president's retweet.

https://twitter.com/nlintheusa/status/935953115249086464?refsrc=email&s=11

The prosecutor's office in the Netherlands that handled the case explained that the fight had happened in May and the boy was sentenced through a juvenile justice program. Source: Saranac Hale Spencer POsted on November 29, 2017

The above President's retweet was one of three anti-Muslim videos that Trump posted on the morning of Nov.29 to his personal Twitter account, @realDonaldTrump, which has more than 40 million followers. All three videos were posted by Jayda Fransen, the deputy leader of Britain First, a far-right group.

The president's retweets also drew condemnation from British Prime Minister Theresa May, who said through her spokeswoman: Britain First seeks to divide communities by their use of hateful narratives that peddle lies and stoke tensions. They cause anxiety to law-abiding people. British people overwhelmingly reject the prejudiced rhetoric pf the far right which is the antithesis of the values this country represents, decency, tolerance and respect."

Native Americans

For centuries, treaties have defined the relationship between many Native American nations and the U.S. More than 370 ratified treaties have helped the U.S. expand its territory and led to many broken promises made to American Indians.

A rare exhibit of such treaties at the Smithsonian's National Museum of the American Indian in Washington, D.C., looks back at this history. It includes one of the first compacts between the U.S. and Native American Nations the Treaty of Canandaigua.

Cherokee Nation

(By the beginning of the nineteenth century, the Cherokee Nation had adopted a written constitution as well as a bilingual newspaper. The new constitutional government fundamentally changed the social structure of the Cherokee from matrilineal to a paternalistic system.

During the 1830's the state of Georgia wanted to expand state jurisdiction to include the Cherokee Nation and moved to do so through a series of legislative actions. The Cherokee Nation opposed these actions through the court system. The United States Supreme Court Upheld Sovereignty through decisions in {Cherokee Nation vs. Georgia and Worchester vs. Georgia}. These two decisions were not supported by the administration! Subsequently, the Removal Act of 1830 was ratified by Congress and signed into law by President Andrew Jackson.

The Cherokee Nation was removed from their traditional territory during the winter and spring of 1838-1839 to Oklahoma. Nearly, 20,000 Cherokees were removed, however, only 16,000

Survived the trip west. The Cherokee Nation reorganized under their original constitution and continue to live in Oklahoma!) Source: Cherokee Nation Web Site – History and Culture Excerpt http//nc-cherokee.com

Reduction of National Monuments

Is the reduction of Bears Ears National Monument by 80 Per Cent and Grand Staircase-Escalante National Monument by 45 Per Cent yet another Land Grab of Native America Land?

Continued Rape of US Turf!

Rape began with clearing of East Coast woodlands and intensified with logging; resulting in gullies over most of the Piedmont and other uplands.

Conservation efforts have slowed deterioration; but the scars remain!

Moving westward the prairies in Kansas and surrounding states, tillage of the Prairies led to the Dustbowl.

Wisdom Ignored

In 1932 Hugh Hammond Bennett toured the High Plains just as the soil began to blow! His diagnosis was it was not natures fault! The cause was man. Indigenous people had farmed for centuries and not lost the soil. In barely a generation Americans had stripped the land of its life-giving layers. Americans had changed the face of the land more than the combined activities of volcanoes, earthquakes, tidal waves, tornadoes, and all the excavations of mankind since the beginning of history.

In 1933 Aldo Leopold invoked "The Conservation Ethic!

For the land to be restored in the manner obvious to XIT cowhands and the Comanches: Grass for animals that eat grass! The struggle went on until Bennett put Roosevelt's agency the Civil Conservation Service (CCC) to work.

CCC

The **Civilian Conservation Corps (CCC)** was a public work relief program that operated from 1933 to 1942 in the United States for unemployed, unmarried men. Originally for young men ages 18–25, it was eventually expanded to ages 17–28.[1]Robert Fechner was the first director of the agency, succeeded by Mcanteen following Fechner's death. The CCC was a major part of President Franklin D. Roosevelt's New Deal that provided unskilled manual labor jobs related to the conservation and development of natural resources in rural lands owned by federal, state, and local governments. The CCC was designed to provide jobs for young men and to relieve families who had difficulty finding jobs during the Great Depression in the United States. Maximum enrollment at any one time was 300,000. Through the course of its nine years in operation, 3 million young men participated in the CCC, which provided them with shelter, clothing, and food, together with a wage of $30 (about $570 in 2017[2]) per month ($25 of which had to be sent home to their families).[3]

The American public made the CCC the most popular of all the New Deal programs.[4]Sources written at the time claimed[5] an individual's enrollment in the CCC led to improved physical condition, heightened morale, and increased employability. The CCC also led to a greater public awareness and appreciation of the outdoors and the nation's natural resources, and the continued need for a carefully planned, comprehensive national program for the protection and development of natural resources.[6]

Enrollees of the CCC planted nearly 3 billion trees to help reforest America; constructed trails, lodges, and related facilities in more than 800 parks nationwide; and upgraded most state parks, updated forest fire fighting methods, and built a network of service buildings and public roadways in remote areas.

Despite its popular support, the CCC was not a permanent agency. It depended on emergency and temporary Congressional legislation and funding to operate. By 1942, with World War II and the draft in operation, the need for work relief declined, and Congress voted to close the program.[10]

SCS

On April 27, 1935 Congress passed Public Law 74-46, in which it recognized that "the wastage of soil and moisture resources on farm, grazing, and forest lands . . . is a menace to the national welfare," and it directed the Secretary of Agriculture to establish the Soil Conservation Service (SCS) as a permanent agency in the USDA. In 1994, Congress changed SCS's name to the Natural Resources Conservation Service (NRCS) to better reflect the broadened scope of the agency's concerns.

The creation of the Soil Conservation Service represented the culmination of the efforts of Hugh Hammond Bennett, "father of Soil Conservation" and the first Chief of SCS, to awaken public concern for the problem of soil erosion. Bennett became aware of the threat posed by the erosion of soils early in his career as a surveyor for the USDA Bureau of Soils. He observed how soil erosion by water and wind reduced the ability of the land to sustain agricultural productivity and to support rural communities who depended on it for their livelihoods. He launched a public crusade of writing and speaking about the soil erosion crisis. His highly influential 1928 publication "Soil Erosion: A National Menace" influenced Congress to create the first federal soil erosion experiment stations in 1929.

WHERE NOW?

Worse

As with many of our follies, we learn nothing from our follies! We now choose to open our National Monuments to Mining and Drilling !

Zoning- Lack Of-Loop Holes-ETC.

Two areas will be discussed! Flood Plains and Coastal **Mountains!**

Flood Plains

Flood Plains appear to have been addressed as part of the National Flood Insurance Reform Act (NIFRA) of 1994 (42 U.S.C. 4101) which appears to allow implementation measures to reduce flood losses, such as elevation, acquisition, or relocation of NFIP-insured structures. What if any restrictions were applied to Casinos built within Levees in Council Bluffs, Iowa? See { pg. 188 of Who Am I Lord? } Zoning maps for the Sacramento River in California were side stepped by building dikes around subdivisions, supposedly preventing flooding of the subdivisions while removing flood plain storage and raising flood levels.

Coastal Mountains

Zoning maps do not exist for coastal areas of California subject to mud flows. Billions have been spent on debris basins, but the loss of lives and homes continue!

For what? To Appease Insatiable Greed!

We have much to learn from the Indigenous Peoples of the world, who know how to live with nature and in carrying out God's Gift of Making Us Caretakers of HIS World!

Clean Coal

""We've ended the war on beautiful, clean coal, and it's just been announced that a second, brand-new coal mine," said Trump, "where they're going to take out clean coal — meaning, they're taking out coal. They're going to clean it — is opening in the state of Pennsylvania, the second one." (Joe Romm, Aug. 23,2017)

Trump considers rolling back rules protecting coal miners from black lung disease

Associated Press

Dec. 16, 2017, 6:04 PM

CHARLESTON, W.Va. (AP) — President Donald Trump's mining regulators are reconsidering rules meant to protect underground miners from breathing coal and rock dust — the cause of black lung — and diesel exhaust, which can cause cancer. An advocate for coal miners said Friday that this sends a "very bad signal."

The Mine Safety and Health Administration has asked for public comments on whether standards "could be improved or made more effective or less burdensome by accommodating advances in technology, innovative techniques, or less costly methods."

Some "requirements that could be streamlined or replaced in frequency" involve coal and rock dust . Others address diesel exhaust , which can have health impacts ranging from headaches and nausea to respiratory disease and cancer.

"Because of the carcinogenic health risk to miners from exposure to diesel exhaust, MSHA is requesting information on approaches that would improve control of diesel particulate matter and diesel exhaust," the agency said.

The Trump administration has said many federal regulations, including pollution restrictions, have depressed the coal industry and other sectors of the economy.

"President Trump made clear the progress his Administration is making in bringing common sense to regulations that hold back job creation and prosperity," Labor Secretary Alexander Acosta said Thursday in releasing his agency's regulatory and deregulatory agenda. "The Department of Labor will continue to protect American workers' interests while limiting the burdens of over-regulation."

The notices on coal dust and underground diesel exhaust had few details. Both were described as "pre-rule stage."

The Cost=Lives

Black Lung

July 1, 20176:00 AM ET

HOWARD BERKES

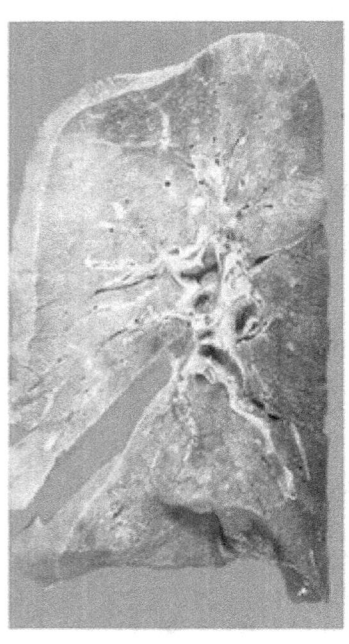 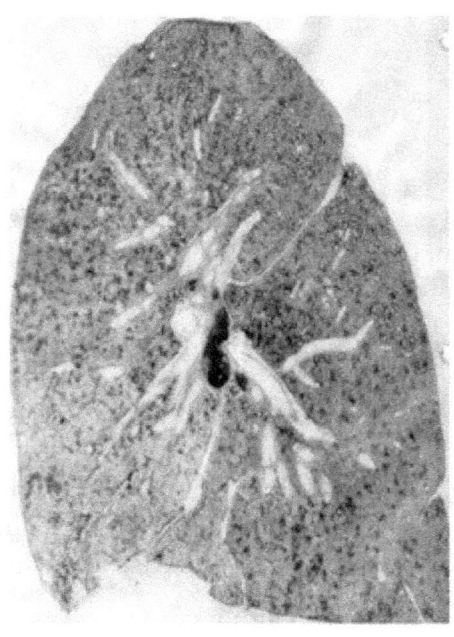 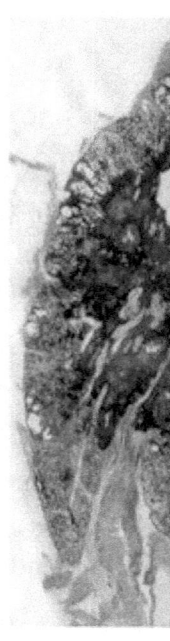

NPR's ongoing investigation of the advanced stage of the fatal lung disease that afflicts coal miners has identified an additional 1,000 cases in Appalachia.

HEALTH

Government Researchers Plan Response To Rising Rates Of Black Lung Disease

The latest NPR count of progressive massive fibrosis, the most serious stage of the disease known as black lung, is nearly 2,000 cases in the region, all of which were diagnosed since 2010.

In the same period, researchers at the National Institute for Occupational Safety and Health reported just 99 cases nationwide. NPR's count is now 20 times what had been considered the official tally of the advanced stage of disease.

NPR contacted black lung clinics, physicians and attorneys across the country. Seventeen in Ohio, Pennsylvania, West Virginia, Virginia and Kentucky provided data. Their diagnoses of advanced disease have not been independently confirmed. But the actual occurrence of disease is likely higher because many clinics across the region and the country were unable to provide data and because others didn't have data for the full 2010-16 time period.

Tax Reform

Impacts of the 2018 Tax Reform remain to be seen: I can only advise everyone to read "The Creature from Jekyll Island by G. Edward Griffin". Pg. 473 excerpt (In practice the Federal Reserve Bank of New York became the fountainhead of the system of twelve regional banks, for New York was the money market of the nation. The other eleven banks were so many expensive mausoleums erected to salve the local pride and quell the Jacksonian fears of the hinterland. Benjamin Strong....president of the Bankers Trust Company [J.P. Morgan] was selected as the first Governor of the New York Reserve Bank. An adept in high finance, Strong for many years manipulated the country's monetary system at the discretion of directors representing the leading New York banks. Under Strong the Reserve System, unsuspected by the nation, was brought into interlocking relations with the Bank of England and the Bank of France.)

January 6, 2018

Good and the Bad

Terminating funding to Pakistan until they seriously attempt to address terrorist groups in their nation is a step in the right direction. Terminating a corrupt government by allowing all people to be represented is more likely to terminate terrorist activity than continued warfare!

The Trump regime continues to make light of Global Warming by carrying snowballs into the White House, etc. as proof that Global Warming is a farce. The administrations push for massive oil drilling expansion confirms **National Greed!** Gone are the days of 55 mph speed limits and funding for solar energy, etc. In the meanwhile, China has expanded their solar energy technology, etc. Highway Deaths continue to rise! **More and more our lifestyle resembles that of the Lemmings!2**

State of The Union 2018 (Annotated by New York Times, Jan, 30,2018)

Mr. Speaker, Mr. Vice President, Members of Congress, the First Lady of the United States, and my fellow Americans:

Less than one year has passed since I first stood at this podium, in this majestic chamber, to speak on behalf of the American People – and to address their concerns, their hopes, and their dreams. That night, our new Administration had already taken swift action. A new tide of optimism was already sweeping across our land.

Each day since, we have gone forward with a clear vision and a righteous mission – to make America great again for all Americans.

Over the last year, we have made incredible progress and achieved extraordinary success. We have faced challenges we expected, and others we could never have imagined. We have shared in the heights of victory and the pains of hardship. We endured floods and fires and storms. But through it all, we have seen the beauty of America's soul, and the steel in America's spine.

Each test has forged new **American heroes** to remind us who we are, and show us what we can be.

We saw the volunteers of the "Cajun Navy," racing to the rescue with their fishing boats to save people in the aftermath of a devastating hurricane.

We saw strangers shielding strangers from a hail of gunfire on the Las Vegas strip.

We heard tales of Americans like Coast Guard Petty Officer Ashlee Leppert, who is here tonight in the gallery with Melania. Ashlee was aboard one of the first helicopters on the scene in Houston during Hurricane Harvey. Through 18 hours of wind and rain, Ashlee braved live power lines and deep water, to help save more than 40 lives. Thank you, Ashlee.

We heard about Americans like firefighter David Dahlberg. He is here with us too. David faced down walls of flame to rescue almost 60 children trapped at a California summer camp threatened by wildfires.

To everyone still recovering in Texas, Florida, Louisiana, Puerto Rico, the Virgin Islands, California, and everywhere else – we are with you, we love you, and we will pull through together.

Some trials over the past year touched this chamber very personally. With us tonight is one of the toughest people ever to serve in this House – a guy who took a bullet, almost died, and was back to work three and a half months later: the legend from Louisiana, Congressman Steve Scalise.

We are incredibly grateful for the heroic efforts of the Capitol Police Officers, the Alexandria Police, and the doctors, nurses, and paramedics who saved his life, and the lives of many others in this room.

In the aftermath of that terrible shooting, we came together, not as Republicans or Democrats, but as representatives of the people. But it is not enough to come together only in times of tragedy. Tonight, I call upon all of us to set aside our differences, to seek out common ground, and to summon the unity we need to deliver for the people we were elected to serve.

Over the last year, the world has seen what we always knew: that no people on Earth are so fearless, or daring, or determined as Americans. If there is a mountain, we climb it. If there is a frontier, we cross it. If there is a challenge, we tame it. If there is an opportunity, we seize it.

So, let us begin tonight by recognizing that the state of our Union is strong because our people are strong.

And together, we are building a safe, strong, and proud America.

Since the election, we have created 2.4 million new jobs, including 200,000 new jobs in manufacturing alone. **After years of wage stagnation, we are finally seeing rising wages.**

Unemployment claims have hit a 45-year low. African-American unemployment stands at the lowest rate ever recorded, and Hispanic American unemployment has also reached the lowest levels in history.

Small business confidence is at an all-time high. The stock market has smashed one record after another, gaining $8 trillion in value. That is great news for Americans' 401k, retirement, pension, and college savings accounts.

And just as I promised the American people from this podium 11 months ago, we enacted the biggest tax cuts and reforms in American history.

Our massive tax cuts provide tremendous relief for the middle class and small businesses.

To lower tax rates for hardworking Americans, we nearly doubled the standard deduction for everyone. Now, the first $24,000 earned by a married couple is completely tax-free. We also doubled the child tax credit.

A typical family of four making $75,000 will see their tax bill reduced by $2,000 – slashing their tax bill in half.

This April will be the last time you ever file under the old broken system – and millions of Americans will have more take-home pay starting next month.

We eliminated an especially cruel tax that fell mostly on Americans making less than $50,000 a year – forcing them to pay tremendous penalties simply because they could not afford government-ordered health plans. **We repealed the core of disastrous Obamacare – the individual mandate is now gone.**

We slashed the business tax rate from 35 percent all the way down to 21 percent, so American companies can compete and win against anyone in the world. These changes alone are estimated to increase average family income by more than $4,000.

Small businesses have also received a massive tax cut, and can now deduct 20 percent of their business income.

Here tonight are Steve Staub and Sandy Keplinger of Staub Manufacturing – a small business in Ohio. They have just finished the best year in their 20-year history. Because of tax reform, they are handing out raises, hiring an additional 14 people, and expanding into the building next door.

One of Staub's employees, Corey Adams, is also with us tonight. Corey is an all-American worker. He supported himself through high school, lost his job during the 2008 recession, and was later hired by Staub, where he trained to become a welder. Like many hardworking Americans, Corey plans to invest his tax-cut raise into his new home and his two daughters' education. Please join me in congratulating Corey.

Since we passed tax cuts, roughly 3 million workers have already gotten tax cut bonuses – many of them thousands of dollars per worker. Apple has just announced it plans to invest a total of $350 billion in America, and hire another 20,000 workers.

This is our new American moment. There has never been a better time to start living the American Dream.

So to every citizen watching at home tonight – no matter where you have been, or where you come from, this is your time. If you work hard, if you believe in yourself, if you believe in America, then you can dream anything, you can be anything, and together, we can achieve anything.

Tonight, I want to talk about what kind of future we are going to have, and what kind of Nation we are going to be. All of us, together, as one team, one people, and one American family.

We all share the same home, the same heart, the same destiny, and the same great American flag.

Together, we are rediscovering the American way.

In America, we know that faith and family, not government and bureaucracy, are the center of the American life. Our motto is "in God we trust."

And we celebrate our police, our military, and our amazing veterans as heroes who deserve our total and unwavering support.

Here tonight is Preston Sharp, a 12-year-old boy from Redding, California, who noticed that veterans' graves were not marked with flags on Veterans Day. He decided to change that, and started a movement that has now placed 40,000 flags at the graves of our great heroes. Preston: a job well done.

Young patriots like Preston teach all of us about our civic duty as Americans. Preston's reverence for those who have served our Nation reminds us why we salute our flag, why we put our hands on our hearts for the pledge of allegiance, **and why we proudly stand for the national anthem**.

Americans love their country. And they deserve a Government that shows them the same love and loyalty in return.

For the last year we have sought to restore the bonds of trust between our citizens and their Government.

Working with the Senate, we are appointing judges who will interpret the Constitution as written, including **a great new Supreme Court Justice,** and more circuit court judges than any new administration in the history of our country.

We are defending our Second Amendment, and have taken historic actions to protect religious liberty.

And we are serving our brave veterans, including giving our veterans choice in their healthcare decisions. Last year, the Congress passed, and I signed, the landmark VA Accountability Act. Since its passage, my Administration has already removed more than 1,500 VA employees who failed to give our veterans the care they deserve – and we are hiring talented people who love our vets as much as we do.

I will not stop until our veterans are properly taken care of, which has been my promise to them from the very beginning of this great journey.

All Americans deserve accountability and respect – and that is what we are giving them. So tonight, I call on the Congress to empower every Cabinet Secretary with the authority to reward good workers – and to remove Federal employees who undermine the public trust or fail the American people.

In our drive to make Washington accountable, **we have eliminated more regulations in our first year than any administration in history.**

We have ended the war on American Energy – and we have ended the war on clean coal. We are now an exporter of energy to the world.

In Detroit, I halted Government mandates that crippled America's autoworkers – so we can get the Motor City revving its engines once again.

Many car companies are now building and expanding plants in the United States – something we have not seen for decades. Chrysler is moving a major plant from Mexico to Michigan; Toyota and Mazda are opening up a plant in Alabama. Soon, plants will be opening up all over the country. This is all news Americans are unaccustomed to hearing – for many years, companies and jobs were only leaving us. But now they are coming back.

Exciting progress is happening every day.

To speed access to breakthrough cures and affordable generic drugs, last year the FDA approved more new and generic drugs and medical devices than ever before in our history.

Donald Trump – Genius-Moron-What?

Definition of GENIUS (MERRIAM WEBSTER)

plural **geniuses** *or* **genii** <u>play</u> \ˈjē-nē-ˌī\
1a *plural* **genii** : an attendant spirit of a person or place
b *plural usually* **genii** : a person who influences another for good or bad
- He has been accused of being his brother's evil *genius*.

2: a strong leaning or inclination : <u>PENCHANT</u>

3a : a peculiar, distinctive, or identifying character or spirit
- the *genius* of our democratic government
b : the associations and traditions of a place
c : a personification or embodiment especially of a quality or condition

4*plural usually* **genii** : <u>SPIRIT</u>, <u>JINNI</u>

5*plural usually* **geniuses**
a : a single strongly marked capacity or aptitude
- had a *genius* for getting along with boys
- —Mary Ross
b : extraordinary intellectual power especially as manifested in creative activity
c : a person endowed with extraordinary mental superiority; *especially* : a person with a very high IQ

Well, one can argue that President Trump qualifies for 3C and 5B. Do we know Trump's IQ?

Dean Keith Simonton published in *Political Psychology* the following table:

The study determined the following IQs of each president as accurate to within five percentage points. In order by presidential term:

PRESIDENT	PARTY	IQ
Franklin Delano Roosevelt	[D]	142
Harry S Truman	[D]	132
Dwight David Eisenhower	[R]	122
John Fitzgerald Kennedy	[D]	174
Lyndon Baines Johnson	[D]	126
Richard Milhous Nixon	[R]	155
Gerald R. Ford	[R]	121
James Earle Carter	[D]	175

Ronald Wilson Reagan	[R]	105
George Herbert Walker Bush	[R]	098
William Jefferson Clinton	[D]	182
George Walker Bush	[R]	091

Simonton used historiometric methods to assess IQ's. An IQ of 160 or higher is generally accepted as Genius. There are no creditable evaluations of Trump's IQ nor proof that he ever took an IQ test. Like his tax records; not available!)

Following a great State of the Union Message, Trump could not resist restating the building of his wall! Trump, we are not in the Ming Dynasty (c. 259-210 B.C.)

If you want a memorial, look towards a Library on your retirement. Yes, security needs to be increased along our southern border! Additional Surveillance towers, surveillance drones, and more guards with dogs will cost less and be more effective!

National Prayer Meeting 2-8-2016

President Trump gave an excellent speech at the breakfast Prayer meeting. He thanked GOD for gifts to mankind, etc.! No one yet has discussed Mankind's obligations to God! GOD appointed Mankind as Caretakers of GODS WORLD! Has Mankind lived up to its responsibility!

One thing is certain, President Trump is a braggart and he would fare better by letting others brag on his accomplishments!

Stock Market 1/18/2018

The major market averages fell modestly in early Thursday trade after Wednesday's solid gains as global investment bank **Morgan Stanley**

Let the Good Times Roll! Future Generations will bear the Consequences!

Book Four is a revival of the series "The World According to Vern". I and my wife Barbara will continue our travels to increase our knowledge of the world and other cultures. New adventures may generate material to be incorporated into new fictional books.

Return to Hawaii February 16,2018

Flew out of Columbia, MO on AA. Weather conditions delayed our arrival in Dallas causing us to miss our flight to Oahu. We also lost our seat assignments causing Barbara to be wedged in between other passengers. An hour into the flight, Barbara passed out and required medical attention. Three volunteer doctors, a nurse and flight attendants cared for her, putting her on an IV and oxygen to overcome dehydration. She was well cared for and improved by the time of landing in Oahu. She was taken off the plane in a wheel chair and her blood sugar content checked prior to being picked up by her son Jimmy and drove to his rental. The rest of our stay went well and the remainder of high points of our visit discussed.

Cono 2003 vs. 2018

In 2003 we were able to visit Cook's monument by renting a Kayak and putting it into the water from a pier and paddling across Cook's bay to the monument. While there, we experienced the best snorkeling

of both trips.

In 2018 kyaking across the bay and snorkeling was only allowed for paid trips.

In 2003 we were able to drive up to active lava flows; in 2018 the active flows had moved further south

and required an 11-mile round trip hike.

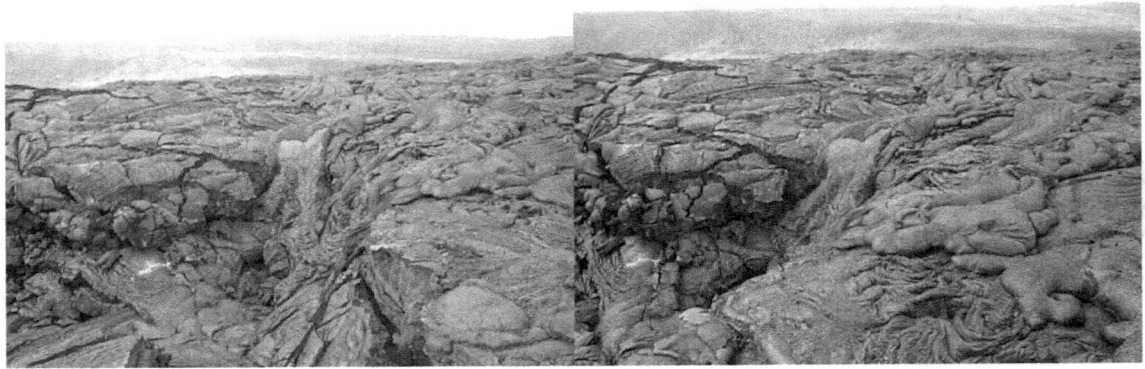

Ohau 2018

Since we had not previously visited Pearl Harbor; we did get to tour the USS Missouri and watch a video of the armistice treaty with Japan!

It was great to spend two weeks with Jimmie as our guide and Barb and Jimmie enjoyed long hikes and cooking together!

Second Amendment of the United States Constitution

"A well regulated Militia, being necessary to the security of a free State, the right of the people to keep and bear Arms, shall not be infringed." Such language has created considerable debate regarding the Amendment's intended scope. On the one hand, some believe that the Amendment's phrase "the right of the

people to keep and bear Arms" creates an individual constitutional right for citizens of the United States. Under this "individual right theory," the United States Constitution restricts legislative bodies from prohibiting firearm possession, or at the very least, the Amendment renders prohibitory and restrictive regulation presumptively unconstitutional. On the other hand, some scholars point to the prefatory

language "a well regulated Militia" to argue that the Framers intended only to restrict Congress from legislating away a state's right to self-defense. Scholars have come to call this theory "the collective rights theory." A collective rights theory of the Second Amendment asserts that citizens do not have an individual right to possess guns and that local, state, and federal legislative bodies therefore possess the authority to regulate firearms without implicating a constitutional right.

Dalai LLAMA Quote From the mouths of Babes come Wisdom It would behoove us to listen to our young people seeing no need for citizens to be armed with military weapons. If they wish to have access to these weapons, they should have access to them on Certified Rifle Ranges and fire them on the rifle range.

P.J. TOBIA: PBS NEWS

There are more than 500 militia groups in the U.S., more than double the number in 2008, according to the Anti-Defamation League. Most of them are right-wing and anti-government.

In addition to the 3 Percent Militia, there's the Oath Keepers, formed in 2009. They're primarily current and former law enforcement and military personnel. Oath Keepers showed up in Ferguson, Missouri, during the protests in the summer of 2015. They said they were there to help keep the peace and protect reporters working for the conspiracy-fueled Web site Info wars.

Meanwhile, thousands have flocked to older groups like the Sovereign Citizens Movement, tax resisters who deny the legitimacy of the American government. Do we want these groups to be armed with military weapons?

The Wall

The **number** of **trucks** crossing the **US-Mexico** border climbed 29 percent from 2009 through 2015, when the figure hit 5.5 million. A number of illegal immigrants have died in trucks due to lack of ventilation! Although X-ray equipment has detected illegals in truck passing through ports of entry. The equipment does malfunction and due to urgency to keep traffic flowing some illegals are not detected and many have died enroute to their destination!

So why is the wall so important when technology exists to expose and monitor illegals! Maintenance of existing equipment, towers with tracking equipment, aerial surveillance -drones - etc.; mounted border guards with dogs, etc.

Nine Eleven Repeat!

The day following nine eleven Saudi Families were flown to Saudi Arabia without being interrogated! The Al Qaida that had carried out destruction were from Saudi Arabia and Egypt!

Now in October 2018 the killing of a reporter by the Saudi's generates statements by Trump that billions of dollars in arm sales may be lost if the US acts against the Saudi government.

Shame On US Should not our mission in the Middle East be to prevent arms build ups? A Coup carried out by Great Britain and the United States destroyed an Iran Democracy that could have been an example for other countries in the Middle East to follow!

.

Readings

A New Geologic Epoch, The Age of Man, pg. 60, National Geographic, March   2011

Antarctic Glaciers Surged After 1995 Ice-Shelf Collapse, John Roach
for National Geographic News, March 6, 2003

Antarctica Gives Mixed Signals on Warming, Bijl P. Trivedi for National Geographic Today, January 25, 2002

Antarctic Ice Collapse Began End of Ice Age? Stefan Lovgren
for National Geographic News, March 17, 2003

Cadillac Desert; The American West and Its Disappearing Water, Revised Edition by Marc Reisner (January 1, 1993)

Geologists Drill into Antarctica and Find Troubling Signs for Ice Sheets' Future New sediment cores from an Antarctic research drilling program suggest that the southernmost continent has had a more dynamic

history than previously suspected by Clay Farris Naff|, Scientific American| April 19, 2010

Good News Bible, American Bible Society, 1977

National Geographic News, Hundreds of Glaciers Melting Faster in Antarctica, Brian Handwerk
June 6, 2007

Ice Core Reveals How Quickly Climate Can Change, Weather patterns can permanently shift in as little as a year, according to the records preserved in an ice core from Greenland, By David Biello, Scientific

American| June 23, 2008

The Worst Hard Time by Timothy Egan 2oo6

Old River Control, US Army Corps of Engineers, Mississippi River Commission and New Orleans District, 1981

100 HEARTBEATS by Jeff Corwin, Rodale Press, 2009

Culture Customs of Afghanistan, Hafizullah, June 30, 2005John Adams: The American President Series: The 2nd President, 1797-1801   John Patrick Diggins and Arthur Mc Schlesinger, June 11, 2003

Griffin, G. Edward 2010, The Creature from Jekyll Island – A Second Look at the Federal Reserve

One Year Chronological Bible, Tynsdale House Publishers, Inc., Carol Stream, Illinois

The Ark of the Covenant, Randall Price, Harvest House Publishers, 200

The Jesus I Never Knew, Philip Yancey, Zondervan, Grand Rapids Michigan, 49530, 1995Truman, David G. Mc Cullough, January 14, 1993

DVD THE 11th HOUR, Produced and Narrated by Leonardo DiCaprio, 2010.

All the Shah's Men, Stephen Kinzer, 2008)

Culture Customs of Afghanistan, Hafizullah, June 30, 2005

John Adams: The American President Series: The 2nd President, 1797-1801, John Patrick Diggins and Arthur Mc Schlesinger, June 11, 2003

One Year Chronological Bible, Tynsdale House Publishers, Inc., Carol Stream, Illinois

The Places In Between, Rory Stewart, May 8, 2006

The Kite Runner, Khaled Hosseini, April 27, 2004

The End Of Days, Zechariah Sitchin, Harper Collins Publisher, 2007

The 12th Planet, Zecharia Sitchin, Harper Collins Publisher, 2007

Three Cups of Tea, Greg Mortenson, Jan. 30, 2007

Truman, David G. Mc Culloch, January 14,1993

All the Shah's Men, Stephen Kinzer, 2008)

Other Books by Vernon Finney

The World According to Vern a series of four;

Three Toes Lives On; The First Environmental Engineers, Ham, Pork and Chop; Oso Gigante, Big Red Mad As A Hatter, Follow The Whales, Berengia to Tierra del Fuego, Who Am I Lord, Chulito and The Leprecaun